# SUPERSCIENCE
# ELECTRICITY

Rob Colson

**W**
FRANKLIN WATTS
LONDON·SYDNEY

First published in 2010 by
Franklin Watts
338 Euston Road
London NW1 3BH

Franklin Watts Australia
Level 17/207 Kent Street
Sydney NSW 2000

Copyright © Franklin Watts 2010

All rights reserved.

Produced for Franklin Watts by
Tall Tree Ltd

Editor: Jon Richards
Designer: Jonathan Vipond
Photographer: Ed Simkins
Consultant: Ade Deane-Pratt

A CIP catalogue record for this book is available from the British Library.

Dewey Classification 537

ISBN 978 0 7496 9522 4

Printed in China

Franklin Watts is a division of Hachette Children's Books, an Hachette UK company.

www.hachette.co.uk

Picture credits:
Cover: main Jerry Horn/Dreamstime.com, tr Vtorous/Dreamstime.com, tm Stuart Key/Dreamstime.com, tl Oriontrail/Dreamstime.com; 1 Fallsview/Dreamstime.com, 3 Tracy Hebden/Dreamstime.com, 4 Bruce Amos/Dreamstime.com, 5b Philip Lange/Dreamstime.com, 6 Fallsview/Dreamstime.com, 7t Levo/Dreamstime.com, 8 Ilja Masik/Dreamstime.com, 9l Les Cunliffe/Dreamstime.com, 9br Ruslanchik/Dreamstime.com, 11t Joris Van Den Heuvel/Dreamstime.com, 12 Vtorous/Dreamstime.com, 13t Aurinko/Dreamstime.com, 14l Tracy Hebden/Dreamstime.com, 14r Mai-Linh Doan/GNU, 15 Dmitry Khochenkov/Dreamstime.com, 16t Anlogin/Dreamstime.com, 17 Krothapalli1/Dreamstime.com, 19b Paddington34/Dreamstime.com, 20 Stevenj/GNU, 21 Sgame/Dreamstime.com, 22 Rasselstein GmbH, 23t Holger Mette/Dreamstime.com, 23b Sergey Eshmetov/Dreamstime.com, 25t NASA, 25b Siemens Pressebild/ GNU, 26 Milan Surkala/Dreamstime.com, 27t Anton Vasilkovsky/Dreamstime.com, 28 istockphoto, 29 Lawrence Wee/Dreamstime.com

Disclaimer:
In preparation of this book, all due care has been exercised with regard to the advice and activities. The publishers regret that they can accept no liability for any loss or injury sustained, and strongly advise adult supervision of activities.

---

**LONDON BOROUGH OF WANDSWORTH**

| 9030 00001 0119 7 | |
|---|---|
| Askews | 27-May-2010 |
| C537 COLS | £12.99 |
| | WWX0006410/0005 |

# ✱Contents

What is electricity?......................4

Static electricity........................6

Current electricity......................8

Electrical circuits.....................10

Carrying electricity................... 12

Resistance ............................ 14

Switches............................... 16

Parallel and series circuits............ 18

Electricity in nature.................... 20

Electromagnetism ................... 22

Generating electricity ................. 24

The cost of electricity................. 26

Using electricity safely ................ 28

Glossary and resources ............... 30

Index ................................. 32

# ✶What is electricity?

Electricity is a source of energy that is all around us. Many of the things we use, from the alarm clock that wakes us up in the morning to the charger that recharges our mobile phone at night, need electricity for their power.

The power lines above this high-speed train carry electricity. The train is connected to the power lines and uses the electricity to power it along.

## Electrical charge

All matter is made from tiny atoms. They are so small that you could fit over a thousand billion of them on a pinhead. Atoms contain even tinier particles called neutrons, protons and electrons. Protons have a positive electrical charge and electrons contain a negative charge. A substance containing more protons than electrons is positively charged. A substance with more electrons than protons is negatively charged. Electrons are the tiniest particles of all, and they can move from one atom to another. The phenomenon we know as electricity is produced by the movement of electrons from negatively charged matter to positively charged matter.

## Using the power

The electricity we use in our homes is made at huge power plants. It is sent from the power plants to wherever it is needed along a network of power lines. Power plants use lots of different energy sources to make electricity. Many of these sources, such as gas and coal, are now running low because we use so much of them. Some power plants also produce harmful waste that pollutes the atmosphere. This is why it is important that we should not waste electricity.

## The history of electricity

The effects produced by electricity have been known for thousands of years. But it was only in the 17th century that scientists discovered the electrical charges that were the cause of these effects. In 1786, Italian doctor Luigi Galvani was examining the dead body of a frog when he accidentally touched an electrically charged scalpel on one of the frog's nerves. Sparks flew and the frog's leg gave a powerful kick. Galvani had discovered that animals use electricity to send signals around their bodies.

Luigi Galvani

This modern power station uses natural gas to produce electricity.

# ✳Static electricity

The crackle when you take off your jumper and the crash of lightning in a storm are both produced by static electricity. Static electricity occurs when two things rub against each other.

## Lightning strikes

When two objects rub against each other, electrons can be knocked off the atoms of one object and stick to the atoms of the other. The object that has lost electrons becomes positively charged, and the object that has gained electrons becomes negatively charged. If the difference in charge, called the voltage, is large enough, electrons move from one object to the other to cancel it out. During thunderstorms, ice crystals in the clouds rub together to produce a positive charge at the top of the cloud and a negative charge at the bottom. Once the difference in charge becomes very large, the electrons at the bottom of the cloud move through the air to the positively charged ground directly below it, causing a spectacular bolt of lightning.

One bolt of lightning contains enough energy to power a small town for a whole year.

This girl's hairs are all positively charged, so they repel each other.

# Static in the home

We all experience static electricity in our everyday lives. Sometimes we receive a small electric shock when we touch a metal door handle. This is because we have built up a negative charge in our bodies by rubbing our shoes on the carpet. Our charged bodies cause a positive charge in the handle, and electrons jump from us to the handle. Charges that are unalike attract each other, while like charges repel. If our bodies build up enough charge, our hairs start to repel each other, and all stand on end.

## Project How to bend water

Give a comb a negative charge by passing it through your hair a few times. Now start a tap running so that a very thin, continuous stream of water flows out of it. Hold the comb very close to the stream of water. The neutrally charged water is given a positive charge when the comb is held near to it. The positively charged water is now attracted to the negatively charged comb, and the stream of water will bend towards the comb.

The water is pulled towards the charged comb.

# Current electricity

Static electricity moves in a single jump. Current electricity is a constant flow of electric charge. This is the form of electricity that we use for our power.

## A flow of energy

When lightning strikes, the electrons move from the cloud to the ground in a fraction of a second. Current electricity is a continuous movement of electrons from a negatively charged point to a positively charged point along a material known as a conductor. The bigger the voltage between the two points, the stronger the current. The electrons do not move all the way along the conductor. They are lined up like a row of marbles, and each one is pushed along by the electron next to it.

Each room of every building in cities such as New York has a wire carrying current electricity to it.

## Batteries

To use the mains electricity, an appliance needs to be plugged in. If we want to carry it around, we need to power it using batteries. Batteries produce a current by converting the chemical energy stored in a cell into electrical energy. Mobile phones are sometimes called 'cell phones' because they run on batteries with cells in them. Once all the energy in the cell is used up, the battery goes flat, and we need to replace it or recharge it by plugging it into the mains.

One end of the battery is positively charged and marked with a + sign. The other end is negatively charged and marked –. Each end must be connected up correctly for the battery to work.

## AC/DC

The electrical current in our homes is an alternating current, or AC. The direction of the current reverses 60 times per second. This is because the generators in power plants that produce the electricity rotate, and every half-rotation of a generator reverses the direction of the current. Most of the appliances in the home, such as lights or televisions, run on an alternating current. Some appliances, such as mobile phones, need special adaptors to change the current from AC to DC, or direct current, in which the current flows only in one direction. The adaptor does not power the phone directly. Instead, it charges up a battery inside the phone.

The battery in a mobile phone is charged with a direct current using an adaptor plugged into the mains.

# *Electrical circuits

For an electrical current to flow, there needs to be a continuous path, called a circuit, for the current to flow through. The current is turned on and off using a switch.

## How do circuits work?

An electrical circuit forms a loop that leads from the power source, such as a battery, to the object being powered, such as a light bulb, and back to the power source. The battery has two connection points: one positively charged, the other negatively charged. The electrons flow from the negatively charged connection, called the anode, to the positively charged connection, called the cathode. The current is said to flow in the opposite direction to the electrons, from the cathode to the anode. A switch can be placed anywhere along the loop. The switch turns the current off by breaking the flow of electrons.

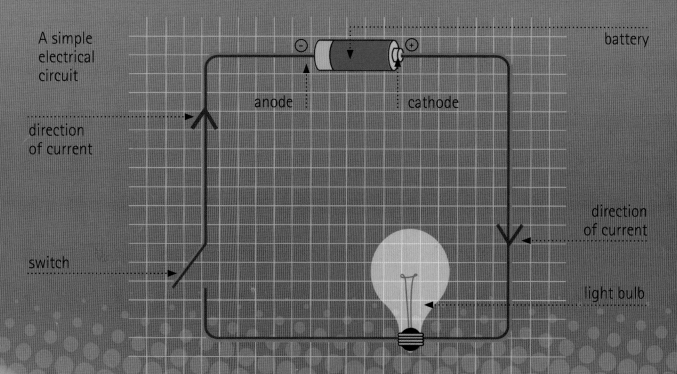

A simple electrical circuit

# Using circuits

Every time you turn on a light in your home, you are completing a circuit that flows from your house to an electricity substation, often many kilometres away. The powerful currents generated by power stations are lowered at substations to make them safe to use. Complicated machines also have smaller switches inside them to control the flow of electricity to each moving part. Computers are by far the most complicated machines that we use. The latest computers contain millions of tiny circuits called microchips. Many other devices in the home, such as televisions and digital clocks, also contain microchips.

There are millions of switches, called transistors, in every tiny microchip.

## Experiment Make a simple circuit

Connect the cathode (+) end of the battery to the bulb with one of the leads. Next connect the bulb to the switch, and finally connect the switch to the anode (−) end of the battery. The bulb will light up when you turn the switch on to complete the circuit, and go out when you break the circuit by turning the switch off. Now connect the switch to the cathode and the bulb, and connect the bulb to the anode. The circuit will work in exactly the same way since the switch can be attached anywhere in the circuit.

Here, the two batteries are in a battery holder, which makes connecting them easier.

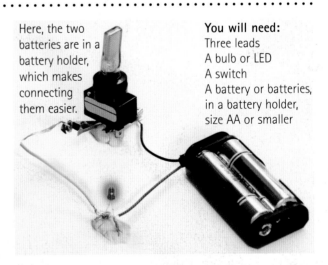

You will need:
Three leads
A bulb or LED
A switch
A battery or batteries, in a battery holder, size AA or smaller

# ✱Carrying electricity

An electrical current is carried around a circuit in a wire. The wire allows the electrons to pass through it and is made from a substance called a conductor.

## Conductors

The electrons in the atoms of conductors are only loosely bound to the rest of the atom, so they move easily from atom to atom when an electrical current is applied to them. Most metals are good conductors. Many electrical cables are made from the metal copper, which is a particularly good conductor. Pure water is a very poor conductor of electricity. However, most water has small amounts of minerals dissolved in it, which turn the water into a good conductor. This is why we should not use electrical appliances in the bathroom unless they are specially adapted.

The wires inside an electrical cable are made from good conductors such as the metal copper.

A series of ceramic disks protects the steel pylon from the electrical cables it supports.

# Insulators

Substances that do not let an electrical current pass easily through them are known as insulators. Most non-metallic substances are insulators. Insulators are used to protect us from the dangerous currents running through electric cables. Pylons support overhead cables that carry very powerful electrical currents. The steel pylons are protected from the current by disks made of ceramic, which is a very good insulator. Luckily for us, air is also a good insulator, so we are safe from the current that flows above our heads.

## Project Testing conductors and insulators

Gather together a selection of objects from around the house. Try to choose things made from lots of different types of material, such as metal, plastic, fabric or clay. Set up the circuit as shown on page 11, but in place of the switch, connect each object up in turn. If the light goes on, the object is a conductor. If the light does not go on, the object is an insulator.

This spoon is made of stainless steel, which is a good conductor so the LED lights up.

# *Resistance

Conductors resist the flow of electrical current, resulting in the loss of energy. This energy is usually lost in the form of heat and light.

The filament in an old-fashioned incandescent light bulb resists some of the current, giving off 95 per cent of the energy as heat and just 5 per cent as light. New energy-saving bulbs work by passing electricity through a gas, and are 80 per cent more efficient.

## Measuring resistance

Good conductors, such as copper, have a low resistance to electrical current. The amount of resistance in a copper cable also depends on the cable's thickness and length. Think of water flowing through a pipe. More water can pass through a wide pipe than a narrow one. In the same way, more electricity can pass through a thick cable than a narrow one. Also, the longer the pipe, the harder it is for the water to travel. Similarly, the longer the cable, the higher its resistance.

## Superconductors

Some materials conduct electricity with no resistance at all at very low temperatures. These are called superconductors. Scientists are hoping to find superconductors that work at higher temperatures. Superconductors could be used to power high-speed magnetic trains. The high-tech research laboratory at CERN in Switzerland uses superconductors at low temperatures to accelerate particles to very nearly the speed of light.

The electric current at the surface of this superconductor (made of liquid nitrogen) acts like a magnet and repels the magnet that floats above it.

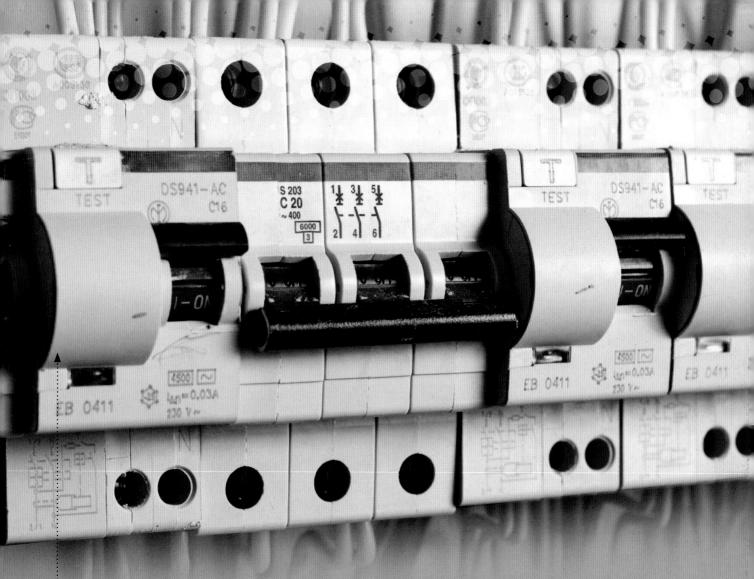

Every home contains a fuse box. The switches turn the power off when there is a dangerous surge. They can also be turned off manually so that an electrician can work on the wires safely.

# Fuses

Electrical cables have low resistance so that they do not waste energy. However, this also means that they can carry dangerously large currents. It is very important that all electrical supplies should be protected in case there is a surge of current, which could damage equipment or even start a fire. For this reason, supplies and plugs are fitted with fuses. A fuse acts like a switch by breaking the circuit to stop the current. Some fuses melt when the current reaches a certain strength. They will need to be replaced before the appliance will work again.

# *Switches

We use switches to turn the electric current on and off. We also use them to change the strength of the current, or to change the current from one circuit to another.

The most common form of switch is known as a toggle switch.

## Making contact

Switches come in many shapes and sizes, but they all do the same thing. They break or complete an electrical circuit by touching or disconnecting two pieces of metal called contacts. When the contacts are touching and electricity can flow, the contacts are said to be closed. When the contacts are separated and not conducting, they are said to be open. Switches can either turn the power on or off, or they can divert the current from one circuit to another.

## Project Make a switch

Strip the insulation from each end of a length of wire and wrap it around each drawing pin. Push one of the drawing pins into the block of wood. Pass the second drawing pin through the eye of the paper clip, and then push it into the block of wood about 1 centimetre from the other drawing pin. Attach the wires to the bulb and battery to complete the circuit. Using the pencil, carefully swivel the paper clip to touch both drawing pins. The light will go on. Remember never to touch the bare electric cables or you'll get an electric shock.

**You will need:**
A small block of wood about 8 centimetres long
A paper clip
Two drawing pins
Three wires
A bulb
A battery
A pencil

## Dimmer switches

Dimmer switches are special kinds of switches that change the strength of the current that flows through a circuit. They are most commonly used to control the brightness of lights. Older dimmer switches reduced the current by increasing the resistance in the switch. This wasted energy in the form of heat. Modern dimmer switches work by turning the power rapidly on and off, thus reducing the amount of energy that reaches the light, but not wasting energy through heat loss.

In this light bank of a theatre, each of the stagelights is controlled by a dimmer switch so that the lighting of the stage can be controlled very precisely.

# *Parallel and series circuits

The components of a circuit can be connected up in three different ways: in series, in parallel, or in a combination of the two.

## Series or parallel?

The circuit on page 10 is an example of the simplest kind of circuit, a series circuit. All the components in a series circuit are connected on a single loop and powered by a single current. In diagram (a), the three lights are connected up in series, and all the lights have to work properly for the circuit to work. If one of the lights is broken, none of the lights will light up. Another way of connecting up components is in parallel, as in diagram (b). Here, the lights are each on their own loop of wire, and the current from the battery is divided up into a separate current for each loop. If one of the lights breaks, the others will stay lit.

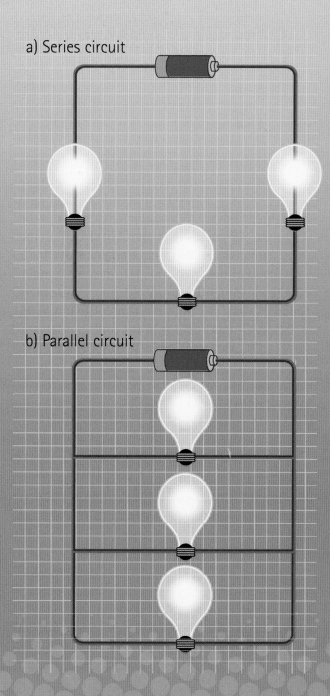

a) Series circuit

b) Parallel circuit

## More power or longer life?

Battery-powered machines, such as toys, remote controls or music players, use batteries connected up in series, in parallel, or both. Using batteries in series means that you are putting more energy into one current, which increases power. Batteries in parallel produce the same amount of power as one battery on its own, but last longer. If you look inside the battery compartment of a remote control, it probably has two batteries side-by-side. They appear to be in parallel, but if you look closely, you will see that the + ends of the batteries are at opposite ends. They are in fact connected in series.

## Project Try it!

Make a parallel circuit using a battery and three LEDs, as shown in (b) opposite. Turn one of the LEDs off by disconnecting it from the circuit, and you will see that the others stay on. Next connect the three bulbs in series as shown in (a). Now if you disconnect one of the LEDs, they all go off. Notice that the LEDs are brighter when they are connected in series than when they are parallel. This means that the battery's power is being used up more quickly, so the brighter lights will go out sooner.

These LEDs are connected in parallel, which means that if one goes out, the others will stay on.

The two batteries in this remote control are connected in series.

Electric eels stun their prey with electric shocks.

# *Electricity in nature

Electricity is found in all living things. Without it we would not be able to move, and we would not be able to think. Some animals even use electricity to kill.

## Electric eels

The electric eel is a scary sight. Up to 2 metres long, it lives in large rivers in South America. It kills its prey by electrocuting it with a wave of electricity powerful enough to knock a horse off its feet. The eel makes electricity in its body using specially adapted organs that turn chemical energy into electricity. The current is conducted by the water and stuns its prey. Scientists are studying the way electric eels make electricity to see whether we can do the same thing to provide power sources for tiny medical implants.

# Electric brains

Our brains are made of billions of tiny nerve cells or neurons. Each neuron is connected to up to 200,000 other neurons at junctions called synapses. When a neuron is activated, or fires, an electrical signal travels from the cell body down to the synapse along a fibre called an axon. This change in charge stimulates special transmitters on the axon terminals, which flow onto the next cell's receptors at the synapse, called dendrites, changing that cell's electrical activity. In this way, each neuron can communicate with thousands of other neurons to make a complex network of cells. Our thoughts are formed by these constantly changing networks.

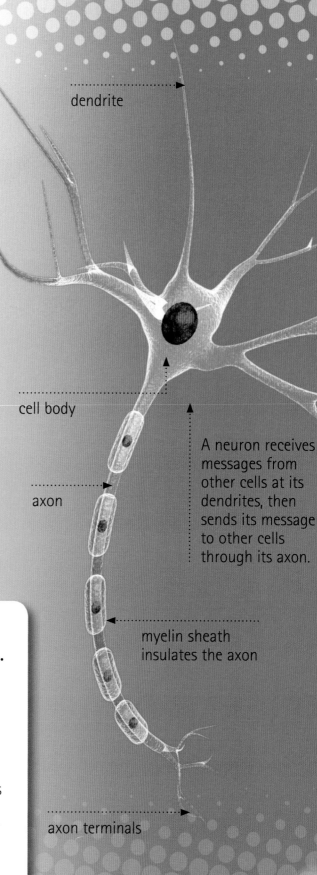

dendrite

cell body

axon

A neuron receives messages from other cells at its dendrites, then sends its message to other cells through its axon.

myelin sheath insulates the axon

axon terminals

## Nerves in the body

Our brains send signals to the rest of the body along nerve cells. The time it takes for the brain to send a signal to your muscles is called the reaction time. To test your reaction times, ask a friend to hold a ruler out in front of you and hold your hands very close either side of the bottom of the ruler. As soon as your friend drops the ruler, try to catch it. You can compare your reaction time with other people's by seeing how close to the bottom of the ruler you catch it.

# ✱Electromagnetism

Every electric current produces a magnetic force around it. We use that force in many types of machine. A magnet that has been made using electricity is called an elecromagnet.

## Giant magnets

Magnetism is a force that acts like electricity. Just as electrical charge is either positive or negative, a magnet has two opposite poles, called north and south. Like poles repel one another and unlike poles attract one another. Some materials, called magnetic materials, are strongly attracted to magnets. Electromagnets strong enough to lift a car can be made by passing a large electrical current through a magnetic material such as iron.

A powerful electromagnet carries a block of crushed tins to a mill to be recycled.

# Electric motors

Electric motors are driven by electromagnetism. A coil of metal is wrapped around the inside of a permanent magnet. When an electrical current is passed through the coil, it generates its own magnetic field, and this magnetic field pushes against the magnetic field in the permanent magnet, causing the coil to rotate. Every half-turn, the direction of the current in the coil is reversed, and this makes it spin continuously. We use the energy in the spinning motor in everything from huge engines to tiny wristwatches.

# Maglev trains

The fastest trains in the world use electromagnetism to move. Maglev trains are suspended a few millimetres above the track using powerful electromagnets. Maglev trains do not have engines. Instead, they are pulled along by the changing magnetic field in the tracks. A maglev train in Japan has reached a top speed of 581 km/h, but if they are run in tunnels with no air to slow them down, they could one day go at more than 6,000 km/h.

In Shanghai, China, a maglev train carries passengers from the airport to the centre of the city. The train covers the 30-kilometre journey in just 7 minutes 20 seconds.

Electric motors come in all sizes but all work in the same way.

cooling fan

permanent magnet

copper coil

# *Generating electricity

Power stations generate electricity using electromagnetism. The process they use can be seen as the reverse of that of an electric motor.

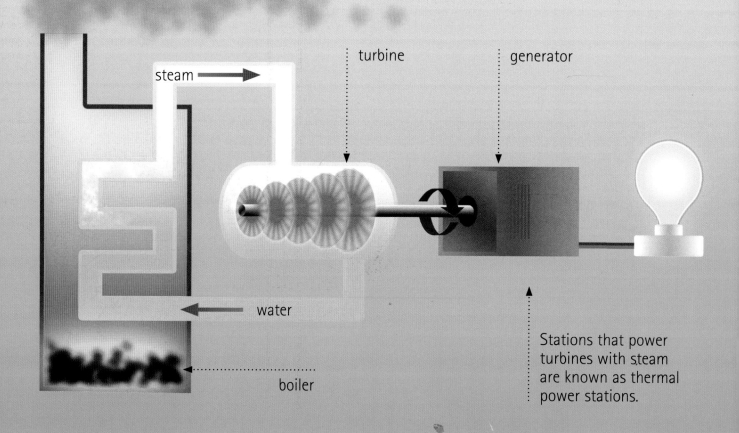

Stations that power turbines with steam are known as thermal power stations.

## Generators

Electricity is produced in a generator. Wire coils spin inside a magnetic field, and this makes an electrical current in the coils. Tiny generators called dynamos can be fitted to bicycles to generate the power to run lights. Small generators such as these use a permanent magnet. Large power stations use much more powerful electromagnets, which means that making electricity uses electricity!

# Using the Sun

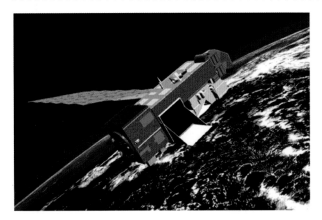

The electronic equipment in a satellite is powered by electricity produced by its solar panels.

Solar panels, also called photovoltaic cells, make electricity directly from sunlight. The panels are made from a material that releases electrons when it absorbs sunlight. These free electrons can then be captured to produce an electrical current. Solar panels were first developed to provide energy for spacecraft. Sunlight in space is very strong, so the panels are an effective way to produce electricity. Solar panels are expensive to make, but they are increasingly also being used down here on Earth.

## Energy sources

Generators in large power stations are powered by huge turbines. Hydroelectric power stations turn the turbines using flowing water. Most other power stations use steam to drive the turbines. The steam is made by heating large quantities of water. The heat is made by burning fossil fuels such as coal, oil or natural gas, or by the chemical reactions inside the atoms in nuclear fuel.

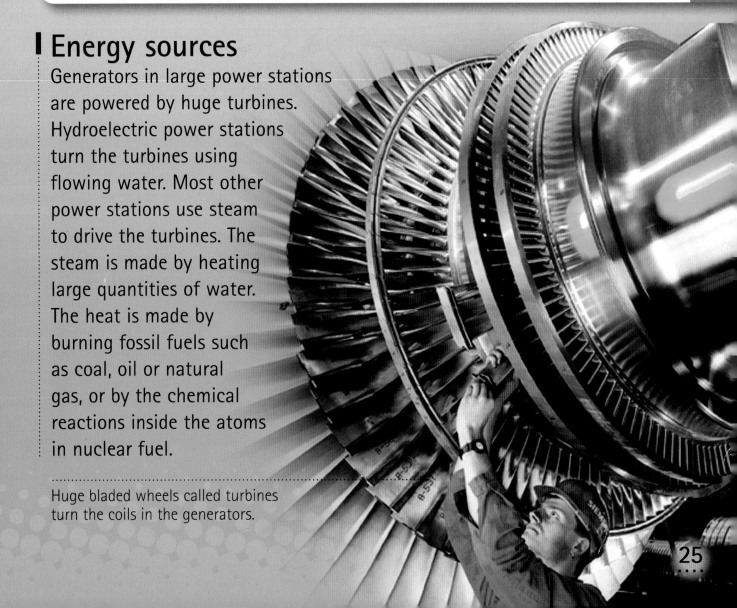

Huge bladed wheels called turbines turn the coils in the generators.

# ✸The cost of electricity

All power stations need an energy source to drive their generators. Some sources of energy are now running low, and many produce dangerous pollution.

## Fossil fuels

Until recently, most of our electricity has been made by burning fossil fuels such as oil, coal and gas. These energy sources are now becoming harder to extract as reserves run low. Burning fossil fuels also releases carbon dioxide into the atmosphere, which causes global warming. Over the coming decades, we will need to replace fossil fuels with less polluting energy sources.

This power station burns coal to produce heat. The fumes from the burning coal are released into the atmosphere.

The cooling towers release excess heat produced by the power plant into the atmosphere by evaporating water.

# Alternative energy

Every source of energy has its own advantages and problems. Nuclear power stations produce small quantities of very dangerous waste. Tidal power stations use the force of the sea's tides for power, but large areas of countryside are flooded when they are built. Wind farms use the power of the wind, but winds do not blow reliably. Geothermal power stations use the heat from the Earth, but can only be built in a few places. In the future, we will need to develop many of these and other sources of energy as alternatives to fossil fuels.

Wind farms use the power of the wind to drive the turbines. This is known as a renewable energy source, because we will never run out of the wind!

## Project Vegetable power

Vegetables have chemical energy in them. You can use the chemical energy in potatoes to make electricity. Push a nail into one end of each potato and a coin into the other end. Using the leads, connect the coin of the first potato to the nail of the other, then connect the remaining coin and nail to the LED to complete the circuit. A chemical reaction in the potatoes between the zinc in the nail and the copper in the coin causes an electrical current, and the LED lights up.

**You will need:**
Two potatoes
Two clean copper coins
Two galvanised nails
Three leads
A small LED

The potatoes are connected in series to increase the size of the current. The more potatoes you add, the brighter the LED will shine.

# *Using electricity safely

Electricity can be very dangerous. A large electric current damages the body's nervous system and may even kill. But sometimes an electric shock can save your life.

## Defibrillators

The heart is a muscle and is controlled by electrical nerve signals. When the signals go wrong, the heart starts to beat irregularly and cannot deliver blood around the body. This can cause death within minutes. In emergencies, doctors use machines called defibrillators to deliver electric shocks of direct current to the heart. The shocks can make the electrical activity in the heart return to normal and save the patient's life.

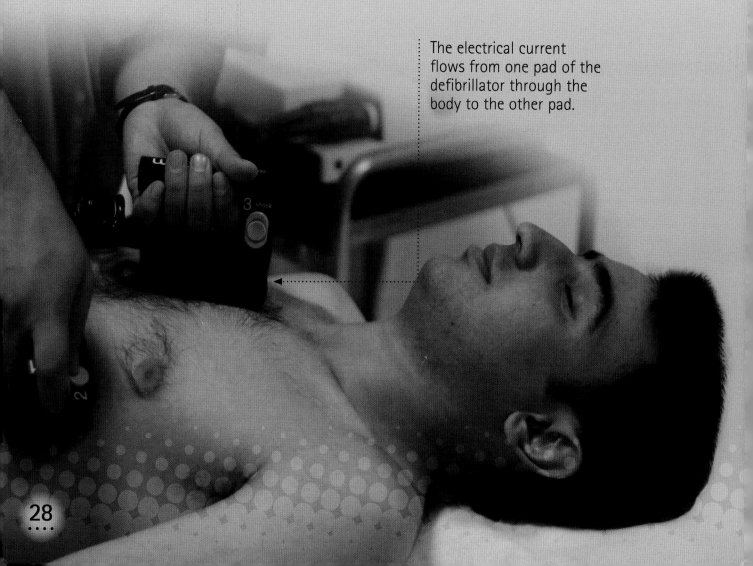

The electrical current flows from one pad of the defibrillator through the body to the other pad.

A lightning rod sticks up from the top of a building so that the lightning strikes it first.

## Safety around the home

Our homes are full of electrical appliances and safety features to protect us in case the electricity goes wrong. Cables and switches are wrapped in insulators, and the power supply is earthed in a similar way to lightning rods. It is very important that everything should be safely insulated and earthed, which is why only a trained electrician is allowed to change the wiring of a house. List all the electrical safety features in your home. Where is the light switch in your bathroom? Why do you think it is there?

## Diverting danger

The electrical charge in a bolt of lightning is enormous. It can destroy buildings, bring down trees and start fires. But don't worry: lightning rarely strikes people. Lightning always takes the shortest route to Earth, so trees and the tops of buildings are most vulnerable. Lightning conductors are fitted to the top of buildings to divert the charge. The lightning strikes the tall rod, and cables from the rod conduct the charge around the building safely into the ground. Conducting a dangerous charge into the ground is called earthing.

# *Glossary

**Adaptor**
A device that changes the alternating current from the mains electricity supply to direct current.

**Cell**
A store of energy inside a battery.

**Charge**
A property of matter that produces electricity. Matter can be positively or negatively charged.

**Chemical energy**
Energy stored in a substance that can be changed into electricity.

**Conductor**
A substance that can easily carry an electrical current.

**Current**
A flow of electrons.

**Electron**
A tiny particle that carries a negative charge.

**Field**
The region around a charged substance that is affected by the force of electricity or magnetism.

**Fossil fuel**
A fuel such as coal or oil that is made from the remains of animals or plants that died millions of years ago.

**Galvanised nail**
An iron or steel nail coated with zinc.

**Generator**
A device that changes mechanical energy into electricity.

**Incandescent light bulb**
A bulb with a filament that resists an electrical current to produce light. Most of the energy is lost as heat.

**Insulator**
A substance that does not easily carry an electrical current.

**LED**
Light Emitting Diode. A low-energy source of light.

**Resistance**
The amount a substance opposes the flow of electrical current through it.

**Turbine**
A wheel in a power station that is turned by steam, flowing water or air and powers the generator.

# *Resources

*The Way Science Works,*
by Kerrod and Holgate (DK, 2008)
Ideas for experiments, plus explanations of scientific theories. Produced in association with the Science Museum.

*Esssential Science: Circuits and Conductors,* by Peter Riley (Franklin Watts, 2006)
A look at the circuits all around us with experiments to see how they work.

*The Real Scientist: Spark!*
by Peter Riley (Franklin Watts, 2008)
Find out how real scientists investigate electricity.

*Do Try This At Home!* by Punk Science (Children's Books, 2008)
Experiment ideas for 'punk scientists', with a DVD of the experiments.

*Richard Hammond's Blast Lab,*
by Richard Hammond (DK, 2009)
The TV presenter shows how to do the experiments he carried out on his science show *Blast Lab*.

*The Horrible Science of Everything,*
by Nick Arnold and Tony De Saulles (Scholastic, 2008)
An exploration of the yucky side of science.

## Websites

www.bbc.co.uk/schools/ks2bitesize
Games, quizzes and fun revision notes on a wide range of science topics.

www.childrensuniversity.manchester.ac.uk
Scientists from the University of Manchester answer questions about electricity and clean sources of energy.

www.scienceprojectideas.co.uk
Ideas for simple projects that you can do at home.

www.sciencemuseum.org.uk
The website of London's Science Museum, which features the latest science discoveries.

www.sciencewithme.com
Games and lots of science project ideas, with worksheets and colouring books to print out. A subscription website that's free to join.

www.explainthatstuff.com/electricity.html
A comprehensive guide to electricity, with project ideas, all on one webpage.

Please note: every effort has been made by the publishers to ensure that these websites contain no inappropriate or offensive material. However, because of the nature of the Internet, it is impossible to guarantee that the contents of these sites will not be altered. We strongly advise that Internet access is supervised by a responsible adult.

# Index

adaptors 9
alternating current 9
alternative energy 27
animals 5, 20
appliances 9, 12, 29
atoms 4, 6, 12

batteries 9, 10, 11, 19
brain 21

cables 12, 13, 14, 15
chemical energy 27
circuits 10–11, 18–19
computers 11
conductors 8, 12, 13, 14
copper 12, 14
current 8–9, 16, 28

defibrillators 28
dimmer switches 17
direct current 9
dynamo 24

earthing 29
electric eel 20
electric motor 23
electric shocks 7, 20, 28
electrical charge 4, 5
electromagnetism 22–3, 24
electrons 4, 6, 8, 10, 12, 25
energy sources 5, 25, 26

fossil fuels 26
fuses 15

Galvani, Luigi 5
generators 9, 24–5
geothermal power 27
global warming 26

heat 14, 17, 26
history 5
home 29
human body 12, 21, 28
hydroelectric power 25

insulators 13, 29

light bulbs 14
lightning 6, 8, 29

machines 11
maglev trains 23
magnetism 22–3
microchips 11
mobile phones 9

National Grid 5
nature 5, 20–21
nerves 21, 28
nuclear power 25, 27

parallel circuit 18, 19
pollution 5, 26
power stations 5, 9, 24–5, 26
pylons 13

reaction times 21
remote control 19
resistance 14–15, 17

safety 28–9
series circuit 11, 18
solar panels 25
static electricity 6–7
substations 11
sunlight 25
superconductors 14
switches 10, 11, 16–17

tidal power 27
trains 4, 14, 23
transistors 11
turbines 25, 27

vegetables 27
voltage 6

water 7, 12
wind farms 27
wires 12